DIESEL ENGINE HISTORY

TERMINOLOGY, FORMULAS, AND CONVERSIONS.

DEVELOPED BY
LARRY R. FOLAND

TABLE OF CONTENTS

HISTORY OF THE DIESEL INDUSTRY

The earliest practical means of obtaining mechanical work by letting an expanding gas move a piston in a cylinder was the steam engine. Its practical execution is generally credited to James Watt The event took place in 1769.

Soon after this date, as indicated in some French and English patents, the idea evolved to produce this expanding gas by burning fuel directly within the working cylinder and doing away with the cumbersome, expansive, and dangerous boiler. Several inventors predicted that this newly conceived internal combustion engine would soon outdate the steam engine.

However, history shows that no practical answer had been found until 1860, when a French mechanic named Lenoir actually went to work and built such an engine,using previously conceived ideas and further improvements. As fuel, he used illuminating gas, which was already commercially manufactured before that time.

The first engine, installed in a work shop in Paris, delivered about one horsepower. The piston sucked a gas-air mixture into the cylinder thru half its stroke.At that moment, the piston exposed a hole or pocket in the cylinder wall in which a continuous electric spark was located. The exploding gas pushed the piston thru the remaining half of the stroke. From this small beginning, an internal combustion engine industry developed as other firms in various European countries took up license.

The working cycle used in this engine was obviously very inefficient. The idea to improve the cycle by compressing the charge and firing it near inner dead center, thereby making use of the entire expansion stroke, did not occur until latter, when Otto in 1864 added the compression stroke. This improved engine efficiency by about 15 to 20%.

The Diesel industry had its inception soon thereafter, when Dr. Rudolf Diesel and others in 1892 proposed to use much higher compression ratios to further improve efficiencies and to make it possible to bum cheap crude oil and even coal dust.

The following is an account of further development, as gathered by Mr. F. P. Gruetzner of Fairbanks Morse Inc., Beloit, Wisconsin. He had been in close touch with the early developments in Germany and was present at some of Diesel's lectures. He also had great share in the development and success of your "Y" engines.

Here is his story:

In 1878, Professor Linde presented a lecture to his students in Munich, Germany, on the thermodynamics, explaining the relatively poor efficiencies of the best steam engines of that time in spite of all the progress that had been made. His students were dutifully impressed and one of them resolved to do something about it by making a note in his notebook. His name was Rudolf Diesel. He turned these thoughts over in his mind for fourteen years and then took out a patent on a heat engine which worked according to the Carnot cycle.

Sadi Carnot,a French artillery officer, published the operation of this cycle in 18 24 in a small pamphlet which was 6uneoin ah rary for several years. -w1ren- it was-re-dis-covered,-it-was- hailed as the most efficient cycle to convert heat into mechanical energy. On this basis, Diesel was going to build the most efficient engine ever conceived. He wanted to compress air in a cylinder to a temperature well above the ignition temperature of the fuel,inject and burn the fuel without a rise in temperature and then expand the gasses to atmospheric pressure.He did not want to use any cooling water jackets. Diesel calculated unheard of efficiencies for it.

In 1893, the Maschinenfabrik Augsburg at Augsburg, and Friedrich Krupp at Essen, Germany, agreed to build and test an engine to Diesel's design. This engine had a 5.9 inch bore,

15.7 5 inch stroke and operated on the 4-cycle principle. It had a crosshead,a combination intake and exhaust valve, injection nozzle, air start valve,safety valve and indicator connection. It had no water jacket. The engine never ran by itself. The piston seized in short order. The longest "run" it ever had was one minute, during which time it made about 8 8 revolutions. It proved only one thing, namely that fuel will ignite without artificial means when injected into highly compressed air. On the basis of this result, Diesel made a triumphant speaking tour through France and sold a license to Carel Freres in Gent, Belgium.

The engine was rebuilt several times with similar results. The piston always seized after a short run, and none of the components of the engine could be tried out under operating conditions. Then Diesel got the idea that the fuel should first be gasified before being injected into the cylinder. He built first an internal, then an external gasifier. In the first one,the fuel passed through a coil located in the combustion chamber and the heat of combustion was to gasify the fuel before it went to the injector. In the second one,the fuel was gasified in a separate boiler and the gas injected into the cylinder by compressed air. With these arrangements,the gas fired very erratically in the cylinder; sometimes the gas burned in the exhaust; sometimes not at all. However, he discovered that a small amount of liquid fuel injected ahead of gas will result in regular firing of the gas in the cylinder. In 189 5, Diesel finally put a water jacket on the cylinder and increased Hie bore to 8.6 7 inches. With this engine, he could at least run long enough to test the different components of the engine.

In 1896, Diesel increased the bore again to 9.84 inches in order to find room for separate intake and exhaust valve,and he added an air compressor for injection air. With this engine, Professor Schroeter tested this engine in 189 7, which he presented in a talk before the V.D.I.and also published in the journal of the society. The following is a summary:

Cylinder Diameter	9.84 Inches
Stroke	14.74.7 Inches
Speed	160 RPM
Piston Speed	393.2.5 Feet Per Min
Slowest Speed	40 RPM
Power Output	19 HP (Approx)
Best Fuel Consumption	0.47 Lbs Per BHP/Hr.
Best Thermal Efficiency	30 Percent

With these results, Diesel had created the most efficient engine of his time. His work was based on, and credit should be given to, Carnot (1824), Brayton (1872), Hargraves (1884), Soehnlein (1884), Koehler (1887), Capitain (1891) and Akroyd Stuart (1891).

Emanual Nobel of St.Petersburg, Russia obtained the patent rights for Russia in 1898. This was considered significant since Russia at that time was the main oil producing country in the world. Many other companies obtained patent rights on this engine at the same time. Among them were some of the best factories of the world - and they all failed to produce a marketable engine. By 1900 all of them had stopped building and the cause of the Diesel engine was at its lowest ebb. The reasons were that the engine was not foolproof for field use and that the manufacturing cost had to be reduced by 30 percent to make it saleable in a competitive market. M.A.N. was the only organization that kept working to make the engine competitive. It is to their credit, especially through their Mr. Lauster' s efforts, that the engine became practical.

In 1901 M.A.N. delivered the first practical engine to Russia; in 1902 the first one to Germany and also the first marine engine to the French Navy. Inspired by the success of M.A.N., Burmeister and Wain came back into the Diesel business in 1903 and completed their first ten engines in 1904. In the same year, the M.A.N. delivered their first engine of 140 HP to the German Navy. 1905 saw the first direct reversible marine engine at M.A.N. In 1906 M.A.N. received an order for four marine engines, of which two were reversible, from the French; they also started building direct reversible two-cycle engines that year. By 1907 the Diesel industry was a profitable and rapidly expanding business and Krupp decided to build again. In 1910, Blohm and Voss at Hamburg started to build double acting two-cycle engines for commercial vessels. The same year M.A.N. submitted a Proposal to the German Navy to build them a 6- cylinder, double acting, two-cycle engine developing 12,000 HP. The first engine was destroyed by fire in 1912. The screws were steam turbine driven. In 1912 the motorship "SELANDIA", built by Burmeister and Wain at Copenhagen, Denmark and equipped with two 4-cycle, 8- cylinder engines developing a total of 2500 HP at 140 RPM, made its first trip to the Far East.

During these years of high activity and rapid development, Diesel was relatively inactive. Wherever he was active, he was unsuccessful. In 1897 he tested a compound Diesel engine consisting of two high pressure and one low pressure cylinders. It was based on the assumption that as long as the compound steam engine is more efficient, the Diesel should be also. The test was a complete failure due to the large heat losses when transferring gas from one cylinder to the other. In the same year, Diesel Motor Fabrik Augsburg was formed, with Diesel participating. In 1898 Diesel made very primitive powdered coal tests, which were given up. In 1906 Sulzer at Winterhur, Switzerland started a Diesel

7

locomotive, with Diesel and Klose assisting. It was finally put into service in 1913 but was not successful. In 1910 Diesel started a factory for small engines, which was soon dissolved. He also built a 30 HP, 800 RPM Diesel engine for automobile use, which was never installed in a car. The world was not ready for Diesel engine in automobiles. In the fall of 1912, Diesel was requested to read a paper on the development of his engine before the German Society of Naval Architects and Marine Engineers. He felt personally attacked by the other speakers and took it much to heart. On the 29th of September, 1913, he disappeared from the channel streamer "DRESDEN", on his way to England. ATew oays lacer fils family found mrntrat payments were due-on- 0cto-ber-1-s-t-w-h-ieh eo-u-lrl- net- be made - his fortune had evaporated and he was bankrupt.

THE INTRODUCTION OF THE DIESEL ENGINE IN THE U.S.A.

Adolphus Bush was sojourning in Germany during 1897 and met Diesel in September. He liked Diesel and his enthusiasm, and wanted to look into his engine. He called his technical adviser, Colonel Meier, to test Diesel's engine and submit a report. Both turned out favorable and on the 9th of October, 1897 Adlolphus Busch bought the rights of the Diesel patents for the U.S.A. and Canada for about $250,000. Busch was a brewer and had no intentions of building the engine himself. He started the Diesel Motor Company of America with the idea to sell licenses to good manufacturers. The first engine was built by the Iron & Machine Co., St. Louis, Missouri according to M.A.N. drawings. It was started in September, 1898 and was finally put in the powerhouse of the Anheuser-Busch brewery in St. Louis for a practical trial where it began to run in January, 1899. Troubles soon developed, but all in all, it did not run badly.

At about that same time, it was realized that the Diesel engine was too expensive to be built and sold in competition with steam engines. The German design called for too much labor which the American market could not absorb, with labor rates as high as $3.00 a day. A.F. Frith of the Diesel Motor Company of America designed a 30 HP engine of simpler design which was delivered to Seib Brothers in Jersey City, New York. It was started during 1900 and gave nothing but trouble. The engine was taken back and replaced with another one which did not work much better. By 1901 the Diesel Motor Company of America was ready to give up. The office was reduced to Colonel Meier and Mr. Reisinger, both working without salary.

In 1902 the American Diesel Engine Company was formed and the International Power Company of New Jersey was to build the engines. By 1903 they already had 27 engines with a total of 1605 HP running and 66 more engines, with 8200 HP, on order. At the World's Fair in St.Louis in 1904 there were three 3-cylinder engines of 225 HP each, direct connected to 150 KW D.C. generators and running in parallel. By 1906, in the United States, there were 108 Diesel engines, with a total of 16,605 HP, running and on order. In 1912 the American patent ran out and several manufacturers took up the building of the Diesel engines. In 1913 a party of three headed by Lt. Niemitz of the U.S. Navy, was visiting German Diesel engine manufacturers and during the following years, they were building the single two-cycle engine for the ship Maumee. By 1913 the number of Diesel engines installed and on order had grown to 83,000 HP.

In spite of this progress, the demand of the American market was for a simpler and more foolproof engine; even if it should be at the expense of fuel consumption. Fairbanks, Morse & CO. therefore concentrated on a hot bulb or semi-Diesel engine, when we entered the field in 1912. We began experimenting with a heavy duty two-cycle horizontal engine of which we put in production a line of 10, 15, 20, and 25 HP engines. In 1914 a vertical two-cycle, crankcase scavenging, stationary engine called the "Y" type was developed. It was followed later by a similar type for marine service called the "C-O" type. Those engines created such a need that they practically dominated their respective markets for years.

One of the improvements to engines consisted of the gradual raising of compression pressure with corresponding reduction of fuel consumption. By 1924, This pressure was so high we obtained "cold start", and we then called it a Diesel engine - our Model 32. It was followed in 1925 by a larger engine of the pump scavenging, two-cycle, type - our Model 33. Both models were developed in Beloit, Wisconsin without the acquisition of foreign patents. Fairbanks, Morse & Co. was firmly launched in the Diesel engine business.

THE HISTORY OF FAIRBANKS MORSE

The company really started in 1830 at St Johnsbury, Vermont, where two enterprising brothers, Erastus and Thaddeus Fairbanks, operated a wagon-making shop and iron foundry. Thaddeus had patented a new plow - which farmers were reluctant to use for fear the iron would poison the soil and discovered a method of cooling now used in basic refrigeration design. The inventive genius of Thaddeus naturally turned toward helping the hemp farmer. He patented machines for hemp dressing equipment. As the farmer had great difficulty in proving the weight of his hemp loads, Thaddeus devised the first "A" level type of scale so loads could be weighed on a scale platform.

This was the founding of the scale industry, and won for the E & T Fairbanks Co. international awards. One of the early agents of the company was Charles Hosmer Morse who started as an apprentice at the age of 17 years, and three years later became a salesman in sales agencies in New York and Chicago. In 1865, Mr. Morse opened an agency in Cincinnati. In 1870, he became a partner in the firm of Fairbanks, Greenleaf & Co. The following year, the Chicago fire wiped out this company, and the morning after the fire Mr. Morse opened a new office under the name of Fairbanks Morse & Co.

Meanwhile, Rev. Leonard Wheeler, Missionary to the Objibway indians, had invented a windmill which was the forerunner of an industry started in 1873 in Beloit, Wisconsin, under the name of Eclipse Windmill Co. One of the earliest and most effective agents of this mill was the firm of Fairbanks Morse & Co. of Chicago. In 1885, the Eclipse Wind Engines Co., as it was then known, bought the plant of the Beloit Wagon Works - at the present site of the Fairbanks Morse & Co. factory - to expand operations. A lift pump was added, and later double acting plunger pumps, bucket pumps, rotary and turbine pumps were included.

A Mr. E.F. Williams was experimenting with a steam engine, and with Mr. Morse, organized the Williams Engine Works. The first internal combustion engine, invented by John Charter was moved to Beloit and production started 1887. By 1890, the plants were consolidated under the Fairbanks, Morse name.

Over in Chicago, a mechanic, George Sheffield by name, living on a farm seven miles away from his work at Three Rivers, Mich. found walking to work irksome. He devised a three wheel hand car he could lift onto the railroad tracks and propel himself to work. By traveling mostly under cover of darkness or in secrecy, he kept this bit of trespassing a secret from the railroad until one evening he was derailed by a broken rail. Realizing this broken rail could wreck a freight train soon due, he borrowed a lantern and flagged the train. The train crew discovered his secret of the rail velocipede and reported it to the Michigan Central officials, who granted him the right to operate his car over their rails. Further investigation of the car resulted in the construction of the first hand cars for railroad use. The Sheffield Velocipede Car Co. was started at Three Rivers. This company along with other railroad equipment, was later acquired by Fairbanks Morse & Co. In 1908, the Three Rivers Electric Co., builders of direct current motors, was added.

Thus, the manufactured products now included scales, internal combustion engines, electric motors, railroad equipment and pumping equipment It was only natural that Mr. Morse became interested in diesel-type engine and the production of semi-diesel engines was started in 1913. From this start, Fairbanks Morse & Co. became the outstanding manufacturer of diesels for municipal, railroad locomotive, marine and industrial use.

In 1944, the Pomona and Westco Pump Companies and facilities were acquired to further complete the lines of pumping equipment.

In 1948, an important improvement was introduced first on the Model 31 diesel engine. It was the "dual fuel" idea adapting the engine to use gas or oil without changing of parts. At that time, the 10 cylinder 3500 HP engine was the largest diesel we had ever built.

In July of 1949, Charles H. Morse, III, then the Vise President of Fairbanks Morse & Co., was killed in a plane crash enroute from Chicago to St. Louis.

The split case double suction pump along with the vertical propeller pumps, were developed in 1954, along with delivery of the first 1000 electric motors for the Automic Energy Commission. The motors were of 1750 and 2000 HP designed and were intended for use at gaseous diffusion plants. About 10 days before the end of the year, Fairbanks Morse & Co. bought the business and equipment of the Armstrong Manufacturing Co. of Milwaukee. Armstrong built rewind starters for small gasoline engines. Fairbanks Morse immediately made plans to manufacture and sell starters through the FM Magneto Division.

The manufacturing of portable truck platforms scales, done in San Francisco and also started in Three Rivers, was transferred to Beloit in December of 1958.

In 1964, George Strichman, Board Chairman and President of Fairbanks, announced a plan to reorganize the parent company to consolidate the ownership of Fairbanks Morse & Co. and to change the corporate name to Colt Industries.

In 1970, F�1rbank"..s Morse acquired the exclusive U.S. license to build the French designed S.E.M.T. Pielstick engine. By 1972, after a successful research and testing program, the Colt- Pielstick PC2 was converted to a dual fuel engine for power generation.

Fairbanks Morse has modified, through design improvements, the Pielstick engine into the Colt- Pielstick engine which is the most powerful "shock qualified" diesel used by the U.S.Navy. This means the Fairbanks Morse version of the Pielstick engine is the only one qualified to propel combat ships.

In the mid 80's our newest line of the Pielstick engine was introduced. The PC4.2 diesel engine is the largest medium speed engine manufactured in the United States. The PC4.2 is a 4 cycle 10 cylinder "V" engine which produces 16,500 BHP at 400 RPM.

ENGINE TERMS AND ABBREVIATIONS

Diesel Engine

An engine in which fuel is ignited entirely by the temperature resulting from the heat of compression of the air supplied for combustion.

Cycle

The completing series of events in each cylinder wherein induction and compression of air, burning of fuel, expansion and expulsion of the working medium are effected in the engine before the series is repeated.

Four-Cycle Engine

An engine completing one cycle in four strokes of the piston or two revolutions of the crankshaft. The cycle events are designated by the following strokes; (1) induction or suction stroke, (2) compression stroke, (3) expansion or power stroke, (4) exhaust stroke.

Single-Acting Engine

One that utilizes the working medium on one end of a single piston.

V-Type Engine

One consisting of two banks of in-line cylinders, arranged along the same crankshaft and with an included angle, usually not over 90 degrees.

Trunk Piston Engine

One in which the connecting rod is attached directly to the wrist pin in the piston. In this type of engine the side thrust caused by the angularity of the connecting rod is taken by that part of the piston called the trunk which bears against the cylinder wall.

Injection System

The system used to introduce at the proper time, and in finely divided form, a metered charge of fuel into the combustion chamber. Combustion chamber (clearance volume) - The volume remaining when the piston reaches the end of the compression stroke.

Turbocharger

An exhaust gas turbine driven centrifugal compressor which utilizes energy in the exhaust gas to increase the density of air supplied to the engine for scavenging and combustion.

Cylinder Pressure Relief Valve

A spring loaded valve which indicates and tends to relieve excessive combustion pressure.

Piston Displacement

The cylinder volume in cubic inches swept by the pistons of the engine. It is equal to the number of cylinders times the area of each piston in square inches times the stroke in inches.

Piston Speed

The total number of feet traveled by a piston in a given time interval, usually expressed in feet per minute.

Horsepower (HP)

A time rate of doing work. One U.S. (and British) horsepower is equal to 33,000 ft. lbs per minute. One horsepower (metric) is equal to 75.0 kilogram meters per second.

Indicated Horsepower (IHP)

The horsepower developed in the cylinder.

Mean Indicated Pressure (MIP)

A defined, constant, hypothetical pressure which would deliver to the top of the piston in one stroke the same work as is actually delivered to the top of the piston by the working fluid in one cylinder.

Brake Horsepower (BHP)

The horsepower delivered by the engine shaft at the output end. The name is derived from the fact that it was originally measured by a brake devise.

Shaft Horsepower

The net power available at the output coupling of a transmission system, such as propulsion gearing, electric propulsion system, slim coupling, etc. It differs from the brake horsepower of the engine by the amount of losses in the system.

Torque

A movement which tends to produce rotation. It is the product of force and radius, expressed in pound-feet or pound-inches.

BHP	Brake Horsepower
BHP-HR	Brake Horsepower - Hour
BMEP	Brake Mean Effective Pressure
BTU	British Thermal Unit
C--	Degree--Centigrade
CFM	Cubic Feet per Minute
F -	Degree Farenheit
FT-LB	Foot-pounds
FPS	Feet per second
GPM	Gallon Per Minute
HHV	High Heating Value

HP	Horsepower
HP-HR	Horsepower per Hour
IHP	Indicated Horsepower
LB/HP/HR	Pounds per Horsepower per Hour
LHV	Low Heating Value
MIP	Mean Indicated Pressure
PSI	Pound per Square Inch
PSIA	Pound per Square Inch Absolute
RPM	Revolutions per Minute
SHP	Shaft Horsepower

Horsepower

One horsepower is the power required to raise a weight of 33,000 pounds 1 foot vertically against gravity in one minute.

Therefore 1 HP equals 33,000 foot pounds per minute.

Torque-Defined

Torque in an internal combustion engine is the twisting effort in pounds exerted on the crankshaft at radius of 1 foot. Torque is therefore expressed in pounds feet.

THIS PAGE INTENTIONALLY LEFT BLANK

FORMULAS AND
CONVERSIONS

THIS PAGE INTENTIONALLY LEIT BLANK

Relation of Torque to Horsepower

Example: Assume a torque of 500 pounds at 1 - foot radius. Assume this torque, or force, constant throughout a complete revolution of the I-foot radius crank arm. Then - 1 foot radius x 2 equals 2 feet diameter x 3.1416 equals 6.2832 feet circumference. 500 pounds x 6.2832 feet equals 3141.6 foot pounds. Assume an engine speed of 1,000 RPM. Then--

$$\frac{3141.6 \times 1000}{33,000} = 95.2 \text{ Horespower}$$

Horsepower Conversion

1 Horsepower delivered for one hour equals 1 horsepower hour.

1 Horsepower equals 2544.1 BTU.

1 Horsepower hour equals .746 kilowatt hour

1 Kilowatt equals 1.34 horsepower.

1 Kilowatt equals 1,000 watts.

1 Watt is the actual electric power transmitted with a current of 1 ampere at a voltage of 1 volt.

Therefore, volt x amperes equals watts.

Approximate Horsepower Developed With Standard Fuels

Gasoline	10 Horsepower hours per gallon
Butane	10 Horsepower hours per gallon
Natural Gas	1/10 Horsepower hour per cubic foot
Diesel Oil	17 Horsepower hours per gallon

Engine Displacement

Engine displacement in cubic inches equals square inches cylinder area x inches stroke x number of cylinders.

BMEP in Pounds (Two Cycle)

$$\frac{\text{HP x 396000}}{\text{Engine Disp. x RPM}} = \text{BMEP}$$

$$\frac{\text{Foot Lbs. Torque x 75.4}}{\text{Engine Displacement}} = \text{BMEP}$$

Torque Equations Where LB FT Load Is Used.

$$\text{T LB FT} \qquad \frac{\text{7 0 4 x Watts}}{\text{RPM}} \qquad \frac{\text{5252 x HP}}{\text{RPM}}$$

Watts	$$\frac{T\,L\,B\,FT \times RMP}{20.4}$$

HP	$$\frac{T\,L\,B\,F\,T \times RPM}{5252}$$

Derivation of Constants:

5252	$$\frac{33,000}{2 \times \text{Inches}}$$

Conversion of Meter Disc Readings to KW

$$\frac{KW\text{-}R x K x M x 3600}{1000 \times T}$$

R	Revolution of meter disc (usually 10)
K	Meter disc constant
M	PT ration x CT ration
T	Exact time in seconds for R revolutions

Data for Engine Test

Total Heat in Fuel Supplied per Hr.

Test - LBS/BHP HR	$$\frac{LBS\ Fuel \times 3600}{\text{Time Sec. x Test BHP}}$$

Test BHP
$$\frac{\text{Total K W Input}}{.746}$$

Test - LBS PER KWH Output
$$\frac{\text{LBS FUEL}}{\frac{\text{K W x (Time Sec)}}{3600}}$$

Test - LBS Per Gallon Fuel
$$\frac{\frac{\text{K W x (Time Sec)}}{3600}}{\text{LBS Fuel LBS Per Gallon}}$$

$$\frac{\text{BTU I L B x LBS x 3600}}{\text{Time E Seconds}}$$

Heat Consumed per BHP Hr. BTU

$$\frac{\text{Total Heat E Fuel Supplied per Hr. BTU}}{\text{BTU / LB x Test LBS / BHP Hr.}}$$

Brake Thermal Eff. of Engine

$$\frac{2545}{\text{BTU / LB x Test LBS / BHP Hr.}}$$

BMEP (FOUR CYCLE)

Brake Mean Effective Pressure is the average cylinder pressure throughout the power stroke of an internal combustion engine, expressed in pounds per square inch.

BMEP

BMEP in pounds

$$\frac{\text{Horsepower x 792,000}}{\text{Engine displacement x RPM}} = \text{BMEP}$$

$$\frac{\text{Foot pound s torque x 150.18}}{\text{Engine displacement}} = \text{BMEP}$$

British Thermal Unit (BTU)

One BTU is the heat required to raise the temperature of 1 pound of water 1-degree Fahrenheit or the amount of heat given off when the temperature of 1 pound of water is lowered 1 degree Fahrenheit.

Formula for Determining Amperes, HP, KW, and KVA

I= Amperes

E= Volts

PF= Power Factor

KW= Kilowatts

KVA= Kilovolts - Amperes

HP= Horsepower

% Eff= Percent Efficiency

⊠ For three-wire, two-phase circuits the current in the common conductor is 1.41 ~times that in either of the other two conductors.

Alternating Current

To Find	Direct Current	Single Phase	Two Phase Four Wire	Three Phase
Amps Where HP is Known	$\dfrac{HP \times 746}{EX\ \%EFF}$	$\dfrac{HP \times 746}{EX\ \%EFF \times PF}$	$\dfrac{HP \times 74}{2 \times Ex\% \times PF}$	$\dfrac{HP}{1.73 \times Ex\% \times PH}$
Amps Where KW is Known	$\dfrac{KW \times 1000}{E}$	$\dfrac{KW \times 1000}{EXPF}$	$\dfrac{KW \times 1000}{2xExPF}$	$\dfrac{KW \times 1000}{1.73 \times ExPF}$
Amps Where KVA is Known		$\dfrac{KVA \times 1000}{E}$	$\dfrac{KVA \times 1000}{2xE}$	$\dfrac{KVA \times 1000}{1.73 \times E}$
Kilowatts	$\dfrac{IXE}{1000}$	$\dfrac{IxExPF \times 1000}{1000}$	$\dfrac{IxFx2xPF}{1000}$	$\dfrac{IxEx1.73 \times PE}{1000}$
KVA		$\dfrac{IXE}{1000}$	$\dfrac{IxEx2 \times 1000}{1000}$	$\dfrac{IxEx1.73}{1000}$
HP (Output)	$\dfrac{IXEX\%Eff}{746}$	$\dfrac{IXEX\% \times PF}{746}$	$\dfrac{IxEx2x\%PF}{746}$	$\dfrac{IxEx1.73x\%PE}{746}$

Diesel Fuel Consumption Calculation

Observe

1. Fuel Consumed
2. Time to Consume Fuel
3. HP Developed During Time of Test

Example

1. 100# Fuel
2. 550 Seconds
3. 1750 BHP

$$\frac{100 \text{ X } 3600}{550} = 654 \text{ lbs. per Hr.}$$

$$\frac{65\ 4 \text{ lbs}}{1750 \text{ BHP}} = .374 \text{ lbs per BHP - } \textit{Specific Fuel Cons}$$

BHP Consumption .374 x 19,350 = 7240 BTU/BHP - Specific BTU

KWH Consumption

$$\frac{.374\ x}{1.34} = 502\ \textit{lbs per } KW$$

KWH Consumption

$$\frac{E\ 'F}{'F}$$

$$\frac{374\ x\ 1.34}{EFF}\ x\ 19,350 = 9700\ \textit{BTU/KW - Specific Fuel BTU}$$

For the Above Example. Fuel

BTU = 19,350/lb.

1 KW = 1.34 BHP

Unit EFF = 100

UNITS OF LENGTH

Basic SI Unit: (Meter) m		
Multiply	**Factor**	**To Obtain**
Inches	0.0254	Meters (m)
Inches	0.254	Decimeters (dm)
Inches	2.54	Centimeters (cm)
Inches	25.4	Milimeters (mm)
Inches	25400	Microns (μ)
Feet	12	Inches
Feet	0.30480	Inches
Yard	3	Feet
Rod	16.5	Feet
Fathom	6	Feet
Miles (Statue)	5280	Feet
Miles (Statue)	1.609344	Kilometers (km)
Miles (nautical)	6076.1033	Feet
Miles (nautical)	1.815996	Kilometers (km)
Miles (nautical)	1.1508	Miles (statue)

UNITS OF AREA

Basic SI Unit: (sq. meter) M^2		
Multiply	**Factor**	**To Obtain**
Square inches (in^2)	0.064516	Square decimeters (dm^2)
Square inches (in^2)	6.4516	Square centimeters (cm^2)
Square feet (fr)	0.092903	Square meters (m^2)
Square feet (fr)	9.2903	Square decimeters (dm^2)
Square feet (ft2)	929.03	Square centimeters (cm^2)
Acre	43560	Square feet (ft2)
Square meters (m^2)	102	Square decimeters (dm^2)
Square meters (m^2)	104	Square centimeters (cm^2)

UNITS OF VOLUME

Multiply	Factor	To Obtain
cubic inches (in^3)	0.016387	Cubic decimeters (dm^3)
cubic inches (in^3)	0.016387	Liters (L)
cubic inches (in^3)	16.387	Cubic decimeters (cm^3)
cubic feet (ft^3)	7.4805194	U.S gallons (gal)
cubic feet (ft^3)	0.02831685	Cubic meters (m^3)
cubic feet (ft^3)	29.31685	Liter (L)
U.S gallons (gal)	231.0	Cubic inches (in^3)
U.S gallons (gal)	0.8331109	Imperial gallons (gal)
U.S gallons (gal)	0.003785412	Cubic meters (m^3)

UNITS OF FLOW

Basic Si Units: m2 (cubic meters/hr) - dm3/sec (cubic decimeters/sec)		
Multiply	**Factor**	**To Obtain**
Cubic feet/minute (ft^3/min)	1.6990	Cubic meters/hour (m^3/hr)
Cubic feet/minute (ft^3/min)	0.41944	Cubic decimeters/second (dm^3/sec)
Cubic feet/minute (ft^3/min)	0.41944	Liters/minute (l/min)
U.S. gallons/minute (gpm)	0.22712	Cubic meters/hours (m^3/hr)
U.S. gallons/minute (gpm)	0.06309	Cubic decimeters/second (dm^3/sec)
U.S. gallons/minute (gpm)	0.6309	Liters/minute (L/min

UNITS OF WEIGHT/MASS

Basic .SI Unit: (kilogram) kg		
Multiply	**Factor**	**To Obtain**
Pound (lb)	16	
Pound (lb)	453.5924	
Pound (lb)	0.4535924	
Short ton (t)	2,000	
Short ton (t)	907.1848	
Long ton (t)	2,240	
Long ton (t)	1,016,047	
Metric ton (t)	2,204.6	
Metric ton (t)	1,000	
Gallon H_2O @ 60 F	28.3378	
Gallon lube oil	7.50	
Gallon diesel fuel oil	7.04	
Gallon heavy fuel oil	7.88 – 8.0	
Cubic feet (ft^3) air @ 68 F, 14.7 PSIA	0.075415	

UNITS OF PRESSURE

Multiply	Factor	To Obtain
Atmosphere	14.69603	pounds/square inch (lb/in^2)
Atmosphere	1.0333	kilograms/square centimeter (kg/cm^2)
Atmosphere	1.01322	Bar
Atmosphere	33.923	Feet water (ft, H_2O) @60° F
Bar	14.50	Pounds/square inch (lb/in^2)
Bar	1.0197162	Kilograms/square centimerer (kg/cm^2)
Bar	102	Kilo Pascal (kPa)
Bar	102	Kilo-newton/square meter (kn/m^2)

UNITS OF TORQUE

Basic SI Unit: (newton-meter) n - m		
Multiply	**Factor**	**To Obtain**
Foot pound (ft lb)	1.355818	Newton-meter (n-m)
Foot pound (ft lb)	7.23301	Kilogram-meter (kg-m)

UNITS OF ENERGY/WORK

Basic SI Unit: (joule) j.		
1 BTU (British thermal unit) will heat 1 lb 1°F.		
i--c-ruurte wrlt---he-at 1 gram water -l €. -· -		
Multiply	**Factor**	**To Obtain**
Foot pound (ft lb)	1.355818	Joule (j)
Foot pound (ft lb)	0.11383	Kilogram-meter (kg-m)
Foot pound (ft lb)	1.28538×10^{-3}	British Thermal Unit (BMU)
Foot pound (ft lb)	3.24×10^{4}	K calorie (Kcal)
Foot pound (ft lb)	3.766×10^{-7}	Watt hour (Wh)
Foot pound (ft lb)	5.050×10^{-7}	Horsepower/hour (hp hr)
British Thermal Unit (BTU)	1055	Joule (U)
British Thermal Unit (BTU)	0.2931	Watt hour (Wh)
British Thermal Unit (BTU)	0.2519957	K calorie (Kcal)
Wait hour	3600	Joule (j)

UNITS OF POWER

The symbols w or kw should be used to define electrical power where losses are involved as with the output of an engine generator set or the input to an electric motor.

Metric horsepower (hp met) is often termed Cheval-Vapeur (ch)		
Multiply	Factor	To Obtain
British horsepower (hp)	0.7456929	Kilowatts **(kw)**
British horsepower (hp)	1.013888	Metric horsepower (hp)
British horsepower (hp)	550	Foot pounds/Second (ft lb/sec)
British horsepower	2545	British thermal units/hour (BTU/hr)
Metric horsepower (hp)	0.735495	Kilowatts (kw)

UNITS OF TEMPERATURE

Water freezes at 0°C or 32°F and boils at 100°C or 212°F at standard atmospheric pressures.

Basic SI Units; (degrees Celsius) °C (degrees Kelvin) °K	
Celsuis to Fahrenheit $$°F = \frac{(°C \times 9)}{5} + 32$$	Fahrenheit to Celsius $$°C = \frac{(°F -32)}{9} \times 5$$
Celsius to Kelvin °K = °C + 273.15	Fahrenheit to Rankin-absolute °R + °F + 459.67

UNITS OF HEATING VALUE OF FUEL OIL

High heat value (HHV) includes the heat liberated by condensation of the vapor produced by combination of hydrogen in the fuel with oxygen in the air during combustion. Such water vapor is not condensed in the diesel cycle so the lower heating value (LHV), which does not include this heat liberation, is normally used in all comutations relative to engine performance.

Basic SI Units: (kilo-joules/kilogram) Kj/Kg		
Btu/lb (British thermal Unit/Pound)	0.5555555	Kcal/kg (Calories/kilogram)
Btu/lb (British thermal Unit/Pound	2.326	Kj/kg (K joules/kilograms)
Kcal/kg (K calories/kilogram)	4.18585	Kj/kg (K joules/kilogram)

UNITS OF SPECIFIC FUEL CONSUMPTION

Coltec Industries, Fairbanks Morse Engine Division normally gives SPC in lb/bhp (British) with the fuels LHV converted to dema Standard 18,190 Btu/lb with parasitic load of specified- engine d-r-i-ven-p-umps-;- The-e-0mmenly u-se-El El:l-F0f)@afl--Stan-d-ru:d LJI-¥-i-S 42,7-00 - kj/kg (18,357.7 Btu/lb) and parasitic load of necessary pumps is not usually included. For proper comparison of SFC's, adjustment must be made to the same LHV standard fuel and to the same included parasitic loads.

A± 5% tolerance is normally allowed on SFC numbers in recongnition of minor differences in indivictual engines of the same design.

Basic SI unit: Grams per kilowatt hour (k/kw hr		
Multiply	By Conversion Factor	To Obtain
lb/bhp/hr (British	447.387	g/bhp/hr (Metric)
lb/bhp hr (British)	608.2274	g/kw hr
g/bhp hr (Metric)	1.35962	g/kw hr

UNITS OF SPECIFIC LUBE OIL CONSUMPTION

Lube oil weight of 7,495 lb/gal (U.S.) is assumed for the following conversation factors.

Multiply	By conversion Factor	To Obtain
Bhp hr/gal (British)	0.000298225	g/bhp hr (Metric)
Bhp hr/gal (British)	0.000405472	g/kw hr
g/bhp hr (Metric)	1.359629	g/kw hr

Note; Lube oil consumption per hour is affected more by engine speed than by engine load. Constant speed engines use nearly as much oil per hour at part load as a full load.

METRIC CONVERSION TABLES

Inches	Decimals	Milimeters	Inches	Decimal	Milimeters
11/64	0.015625	0.3969	13/32	0/40625	10.3187
1/32	0.03125	0.7937	27/64	0./421875	10.7156
3/64	0.046875	1.1906	7/16	0.4375	11.1125
1/16	0.0625	1.5875	29/64	0.453125	11.5094
5/64	0.078125	1.9844	15/32	0.46875	11.9062
3/32	0.09375	2.3812	31/64	0.48375	12.3031
11/64	0.109375	2.7781	½	0.5	12.7000
1/8	0.125	3.1750	33/64	0.515625	13.0969
9/64	0.140625	3.5719	17/32	0.53125	13.4937
5/32	0.15625	3.9687	35/64	0.546875	13.8906
111/64	0.171875	4.3656	9/16	0.5625	14.2875
1/6	0.1875	4.7625	37/64	0.578125	14.6844
13/64	0.203125	5.1594	19/32	0.59375	15.0812
7/32	0.21875	5.5562	39/64	0.609375	15.4781
15/64	0.234375	5.9531	5/8	0.625	15.8750
¼	0.25	6.3500	41/64	0.640625	16.2719
17/64	0.265625	6.7469	21/32	0.65625	16.6687
9/32	0.28125	7.1437	43/64	0.671875	17.0656
19/64	0.296875	7.5406	11/16	0.6875	17.4625
5/16	0.3125	7.9375	45/64	0.703125	17.8594
21/64	0.328125	8.3344	23/32	0.71875	18.2562
11/32	0.34375	8.7312	47/64	0.734375	18.6531
23/64	0.359375	9.1281	¾	0.75	19.0500
31s	0.375	9.5250	49/64	0.765625	19.4469
25/64	0.390625	0.9219	25/32	0.78125	19.8437
51/64	0.796875	20.2406	29/32	0.90625	23.0187
1-37/16	0.8125	20.637	29/64	0.92189	23.4111

53/64	0.828125	21.0344	15/16	0.9375	23.8125
27/32	0.84375	21.4312	61/64	0.953125	24.2094
55/64	0.859375	21.8281	31/32	0.96875	24.6062
7/8	0.875	22.2250	63/64	0.984375	25.0031
57/64	0.890625	22.6219			

Metric Conversion Tables; Millimeters to Inches/Inches to Millimeters.

mm	Inches	Inches	mm	mm	Inches	Inches	mm
0.01	0.00039	0.001	0.0254	0.7	0.02756	0.07	1.778
0.02	0.00079	0.002	0.0508	0.8	0.03150	0.08	2.032
0.03	0.00118	0.003	0.0762	0.9	0.03543	0.09	2.286
0.04	0.00157	0.004	0.1016	1	0.03937	0.1	2.54
0.05	0.00197	0.005	0.1270	2	0.07874	0.2	5.08
0.06	0.00236	0.006	0.1524	3	0.11811	0.3	7.62
0.07	0.00276	0.007	0.1778	4	0.15748	0.4	10.16
0.08	0.00315	0.008	0.2032	5	0.19685	0.5	12.70
0.09	0.00354	0.009	0.2286	6	0.23622	0.6	15.24
0.1	0.00394	0.01	0.254	7	0.27559	0.7	17.78
0.2	0.00787	0.02	0.508	8	0.31496	0.8	20.32
0.3	0.01181	0.03	0.762	9	0.35433	0.9	22.86
0.4	0.01575	0.04	1.016	10	0.39370	1	25.4
0.5	0.01969	0.05	1.270	11	0.43307	2	50.8
0.6	0.2362	0.06	1.524	12	0.47244	3	76.2

mm	Inches	Inches	mm	mm	Inches	Inches	mm
13	0.51181	4	101.6	30	1.18110	21	533.4
14	0.55118	5	127.0	31	1.22047	22	558.8

15	0.59055	6	152.4	32	1.25984	23	584.2
16	0.62992	7	177.8	33	1.29921	24	609.6
17	0.66929	8	203.2	34	1.33858	25	635.0
18	0.70866	9	228.6	35	1.37795	26	660.4
19	0.74803	10	254.0	36	1.41732	27	685.8
20	0.78740	11	279.4	37	1.4567	28	711.2
21	0.82677	12	304.8	38	1.4961	29	736.6
22	0.86614	13	330.2	39	1.5354	30	762.0
23	0.90551	14	355.6	40	1.5748	31	787.4
24	0.94488	15	381.0	41	1.6142	32	812.8
25	0.98425	16	406.4	42	1.6535	33	838.2
26	1.02362	17	431.8	43	1.6929	34	863.6
27	1.06299	18	457.2	44	1.7323	35	889.0
28	1.10236	19	482.6	45	1.7717	36	914.4
29	1.14173	20	508.0				

$$F° = \frac{C°}{S} \times 9 + 32 \qquad C° = \frac{F° - 32}{9} \text{ :c S}$$

F°	C°	F°	C°	F°	C°	F°	C°	F°	C°
-160	-106.67	150	65.56	370	187.78	590	310.00	810	432.22
-140	-95.56	160	71.11	380	193.33	600	315.56	820	437.78
120	48.89	170	76.67	390	198.89	610	321.11	830	443.33
-100	-73.33	180	82.22	400	204.44	620	326.67	840	448.89
-80	-62.22	190	87.78	410	210.00	630	332.22	850	454.44
-60	-51.11	200	93.33	420	215.56	640	337.78	860	460.00
-40	-40.00	210	98.89	430	221.11	650	343.33	870	465.56
-20	-28.89	220	104.44	440	226.67	660	348.89	880	471.11
0	-17.78	230	110.00	450	232.22	670	354.44	890	476.67
20	-6.67	240	115.56	460	237.78	680	360.00	900	482.22
32	0.00	250	121.11	470	243.33	690	365.56	910	487.78
40	4.44	260	126.67	480	248.89	700	371.11	920	493.33
50	10.00	270	132.22	490	254.44	710	376.67	930	498.89
60	15.56	280	137.78	500	260.00	720	382.22	940	504.44
70	21.11	290	143.33	510	265.56	730	387.78	950	510.00
80	26.67	300	148.89	520	271.11	740	393.33	960	515.56
90	32.22	310	154.44	530	276.67	750	398.89	970	521.11
100	37.78	320	160.00	540	282.22	760	404.44	980	526.67
110	43.33	330	165.56	550	287.78	770	410.00	990	532.22
120	48.89	340	171.11	560	293.33	780	415.56	1000	537.78
130	54.44	350	176.67	570	298.89	790	421.11	1010	543.33
140	60.00	360	182.22	580	304.44	800	426.67	1020	548.89

F°	C°	F°	C°	F°	C°	F°	C°	F°	C°
1030	554.44	1240	671.11	1450	787.78	1660	904.44	1870	1,021.11
1040	560.00	1250	676.67	1460	793.33	1670	910.00	1880	1,026.67
1050	565.56	1260	682.22	1470	798.89	1680	915.56	1890	1,032.22
1060	571.11	1270	687.78	1480	804.44	1690	921.11	1900	1,037.78
1070	576.67	1280	693.33	1490	810.00	1700	926.67	1910	1,043.33
1080	582.22	1290	698.89	1500	815.56	1710	932.22	1920	1,048.89
1090	587.78	1300	704.44	1510	821.11	1720	937.78	1930	1,054.44
1100	593.33	1310	710.00	1520	826.67	1730	943.33	1940	1,060.00
1110	598.89	1320	715.56	1530	832.22	1740	948.89	1950	1,065.56
1120	604.44	1330	721.11	1540	837.78	1750	954.44	1960	1,071.11
1130	610.00	1340	726.67	1550	843.33	1760	960.00	1970	1,076.67
1140	615.56	1350	732.22	1560	848.89	1770	965.56	1980	1,082.22
1150	621.11	1360	737.78	1570	854.44	1780	971.11	1990	1,087.78
1160	626.67	1370	743.33	1580	860.00	1790	976.67	2000	1,093.33
1170	632.22	1380	748.89	1590	865.56	1800	982.22	2010	1,098.89
1180	637.78	1390	754.44	1600	871.11	1810	987.78	2020	1,104.44
1190	643.33	1400	760.00	1610	876.67	1820	993.33	2030	1,110.00
1200	648.89	1410	765.56	1620	882.22	1830	998.89	2040	1,115.56
1210	654.44	1420	771.11	1630	887.78	1840	1,004.44	2050	1,121.11
1220	660.00	1430	776.67	1640	893.33	1850	1,010.00	2060	1,126.67
1230	665.56	1440	782.22	1650	898.89	1860	1,015.56	2070	1,132.22

THE DIESEL PRINCIPLE

The diagram shown in Figure 1 is called a PV or pressure volume diagram. It is also called an indicator diagram, because it records through an indicator the change in pressure as the piston moves from 0.0.C. to I.D.C. and back.

Below the diagram is shown a cylinder with the piston in these two positions.

First, let us have a clear understanding of the ten-compression ratio. It is merely the ratio of the total cylinder volume (VS VC) when the piston is in O.D.C. to the combustion chamber volume (VC) when the piston is in·1.o.c.

The heavy l.ine shmvs a typical Diesel engine diagram. The lower curve indicates the rise in compression pressure, and the upper curve, the fall 1n pressure upon expansion of the gasses as the piston is being pushed back to 0.0.C. (1).

Diesel, in his original concept, intended to realize as far as possible the Carnot cycle, which is the most efficient one for converting heat energy into mechanical work. The basic requirements for this cycle are, to burn all the fuel at the highest possible temperature, and to convert the heat into mechanical work dm-1n to the lowest possible temperature. Diesel proposed to accomplish this in the following way: (See the thin line extension of the Diesel diagram).

Beginning at (1), compression of air takes place at constant temperature to (2), and pressure rises very little (isothennal). This meant that compression heat had to be removed. He intended to do this by spraying water into the heated air. From (2) compression proceeds to point (X) without taking away or adding any heat (adiabatic). At point (X), the highest possible pressure and temperature is to be established by compression heat only, not by the burning of the fuel. Then, as the piston recedes, heat from the burning fuel is gradually added so that the temperature remains constant as it was at (X) (isothennal) all the way down to (3) where the injection is stopped. From·there, the gas expands without adding or subtracting any heat, pushing the piston back to (1) (adiabatic).

The highest theoretical efficiency for this cycle is 100%, but the pressure and temperature at (X) would have to be infinitely high.

Since this was unattainable, Diesel revised his calculations to a maximum compression pressure at (X) of around 1400 PSI, for which the theoretical efficiency would still be 73%.

However, the difficulties were still unsur- mountable and Diesel did not realize the Carnot cycle. After further com- promise, it settled down to point (A) where the compression ratio is only around 15:1, and the theoretical efficiency is down to 57%. The compression pressure at this point is around 550 PSI and that is about what we are still using at the present time.

After various losses such as heat, friction, and other parasitic losses were subtracted, the first practical diesel engines had an overall efficiency of slightly over 30%.

This was, however, far better than could be reached before with the low com- pression Otto engine. FurtheMore, it cleared the way for building large engines which would burn the cheap petroleum oils at good efficiency.

Caning back again to point (A) on the diagram, we find that the temperature at this point (about 900 degrees) is still high enough above the ignition temperature of the fuel, and separate ignition means are not required even for starting the engine. To obtain the highest possible efficiency, however, the fuel injection, must begin sight sudden pressure rise to a peak (B). Many volumes have been written about the curve between (A) and (B), exploring the possibility to reduce the sharp rise. But in spite of much development, we still have this peak, especially on the higher speed engines.

Invariably, we find that earlier injection will produce better fuel consumption, but at the same time. results in higher peak pressures.

To keep the resulting mechanical loads on the affected engine parts, especially piston rings, to a reasonable limit, we are restricting the peak on our large slow speed engines to 850 PSI and our O.P. engines to between 1100 and 1200 PSI.

This can be done by proper injection timing according to the applicable instruction books, which must be carefully followed.

One way to reduce the peak pressures is by means of a precombustion chamber. However, this always entails a considerable loss in efficiency and is, there- fore, not applied to larger engines above 6" bore.

In the early days of the diesel engine, the fuel was injected with the help of an air blast, requiring an engine-attached air pump with a pressure output higher than the compression pressure.

Later, it was found that the fuel could be atomized good enough without the air blast. This was then named the solid injection and is commonly used today.

The elimination of the air pump resulted not only in a cost saving but also in a gain in efficiency by reducing the engine's mechanical losses.

This, together 'with many other refinements, has advanced the overall 11 thenna 1 efficiency of a modern engine to around 35% and by addition of turbocharging to around 38% at full load.

The area within the diagram represents the work done. When comparing the three cycles, the Otto, the. Diesel and the Carnot, you will notice that the area becomes larger as the compression ratio increases. This is also a direct sign that the efficiency increases with higher compression ratios. The upper limits are imposed by practical considerations, such as the enormous pressure and temperature and corresponding friction and heat losses.

From the area of the diagram, we can calculate an average of all the pressures. This is called the mean indicated pressure (M.I.P.). When we multiply this factor with the mechanical efficiency of the engine, then we have the brake mean effective pressure (B.M.E.P.). The mechanical efficiency is usually around .8 and comprises all mechanical losses, such as friction and the power to drive various pumps and the blower.

The Diesel diagram looks basically the same for the two-cycle or four-cycle engine except that the latter has two extra strokes, one for air intake and one for pushing out the exhaust gases.

On the two-cycle engine, these two functions are accomplished at the end of the expansion stroke and the beginning of the compression stroke.

Therefore, the two-cycle requires only one revolution (per cylinder) for each power stroke, while the four-cycle requires two revolutions.

This means that the four-cycle, to be equivalent in power output for a given cylinder displacement, has to burn roughly twice as much fuel in each power stroke than the two-cycle.

This will beccme increasingly difficult for the four-cycle engine because the two-cycle is now also being turbocharged.

The taking of pressure volume indicator cards has been discouraged on Diesel engines, primarily for the reason that the sudden pressure changes affect the correctness of the readings. Better indicators were needed and the electrical one developed by the M.I.T. was used for the diagrams on SSNS, which we will use for an example. These diagrams were taken on a 10-cylinder 8-1/8 O.P. engine running at 720 RPM and rated 160 HP per cylinder.

On the M.I.T. indicator, the compression line appears on the L.H. side of I.D.C.

The cycle begins with the opening of the scavenging air ports 40 degrees before lower dead center. At that point, the pressure in the cylinder is still well below the air receiver pressure which 'will cause a short blm·, back. Then the pressure drops rapidly, and as soon as it is below the receiver pressure, scavenging begins. Due to the momentum of the gases, the pressure· dips momen-tarily below atmospheric pressure and then scavenging continues in a 'il avy line. When the exhaust ports close 56 degrees after top dead center, the pressure in the cylinder has reached somewhat over 2 PSI, and at the close of the air intake ports 64 degrees after dead center 2.5 PSI. This is above the density of the surrounding atmosphere and puts the engine into the supercharged class according to "Standard Practices for Stationary Engines". The temperature at this point is estimated at about 200 degrees F..

Compression begins here and continues in the conventional manner until injection of the fuel begins. By timing of the pumps, injection lasts from 14 degrees before to 6 degrees after inner dead center on the engine on which this investment was made.

Since there is a delay between the timing and the actual opening of the injection valve, an indicator was attached to the needle showing its movement. Added to our general picture, it shows a delay of 6 degrees for the beginning of the actual injection. During the first part of injection, the fuel has to_be vaporized which takes heat from the gases in the cylinder. This is indicated by a small irregularity in the compression line towards the end of the stroke. Toe compression end pressure in the diagram is 575 PSI. The temperature at this pressure is about 800 degrees to 900 degrees F.

As soon as the first gasified fuel begins to burn, the pressure rises rapidly to an amount which, on present engines, is limited to 1150 PSI maximum. Combustion continues after maximum pressure, and a maximum temperature of about 2500 degrees F. is reached after expansion has begun. Then expansion proceeds adiabatically until the exhaust ports open. This happens 56 degrees before outer dead center. The pressure drops rapidly until the air receiver pressure the remaining gases out of the cylinder. This completes the cycle of the Dies-21 engine.

The diagram also shows the pressure in the air receiver which showed a gauge pressure of 3 PSJ with maximum and minimum lines as indicated. It also shows the pressure in the exhaust manifold which \.1 as an average of 5 inches of water and had waves as indicated.

The above cycle is repeated with variations on all Diesel engines. We will describe sane of the more important ones.

The "crankcase scavenging engine" is a single-piston two-cycle type. As the piston moves up on the compression stroke, the underside of the piston sucks air into the crankcase. On the power stroke, this air is compressed slightly. Near the bottom of the exhaust ports, located on the side of the cylinder, open first to relieve the pressure. Then the air intake ports, located at the opposite side, open and admit the air from the crankcase. Due to the low volumetric efficiency of the crankcase as a pump, the amount of fresh air for the cylinder is limited, and the obtainable mean effective pressure is low. The fuel consumption of these engines, however, is good.

The "single acting, pump scavenging, two cycle" type has one piston per cylinder in the conventional manner. The simplest form has air intake and exhaust ports near the end of the stroke. It has a separate scavenge air pump which supplies air in excess of the piston displacement. It gives higher MEP and is suitable for larger powers. A typical example is our 16 x 20 Model 33. The scavenge pump may be direct driven from the engine-driven from an auxiliary unit or the air could be supplied by a motor driven blower. Some engines have a combination of valves and ports. The EMO engine uses a row of ports at the outer dead center of the piston for the entry of the scavenge air and a set of exhaust valves in the cylinder head. The process has been reversed by other builders who use air intake valves in the cylinder head and exhaust ports in the cylinder.

For the very largest powers, sane manufacturers are building a "double acting two cycle" engine. These engines must have a crosshead, piston rod and stuffing box for each cylinder which makes them rather tall. They do not produce twice the power of the single-acting type since the piston rod reduces the area of the lower side and also causes additional cooling. The best known one is the one manufactured by the M.A.N. It has two rows of exhaust ports, one for the upper and one for the lower side, and one row of scavenge air ports serving both sides. The air ports are in the middle of the cylinder, the exhaust ports are directly above and below them. Rotary valves are installed in the exhaust lines to close the passage during the end of the scavenging period for supercharging. The double-acting two-cycle is more complicated and did not get a good hold in the United States. Manufacturers prefer to build even the largest sizes in the single acting type.

The "four cycle" type was used on internal combustion engines long before Diesel. Due to the clearness of its operation, it was taken over by the Diesel industry at the start. It has slowly and constantly been improved and is still enjoying an active life.

Two revolutions are needed to complete the cycle.   On account of two strokes with hardly any load, the suction and exhaust one, more time is available for cooling and pistons for four cycle engines of·large cylinder diameter can operate without special piston cooling. For the same reason, the bearings have more time to dissipate the heat. For the same speed and power, however, the engine is taking up more space and is heavier.

The "supercharger four-cylce" engine has been developed. In this type, some of the energy left in the exhaust is used for driving an exhaust turbine, which is directly connected to an air compressor. This compressor charges the cylinder with air of a density greater than the density of the surrounding atmosphere. It is therefore able to burn more fuel and produce more power in a given cylinder size. This results in a reduction of weight and space per horsepower. It also affects favorably the fuel consumption at partial loads.

During the time when the double acting two· cycle engine was in the development stage, the double acting principle was tried on the four-cycle engine. Engines of this type became large, heavy and cumbersome and enjoyed only a short span of life.

Another combination tried on the four-cycle engine was to use the labor side of the piston as a scavenge pump. Theoretically, such an engine would get almost twice as much air as an atmospherically charged one and should deliver more power. It meant employing a crosshead, piston rod, lower cylinder head with air inlet and discharge valves and an air receiver. The engine became heavy and expensive and as so often experienced, the gain in power was not worth the complications and expenses gone to. This design, together with the double-acting four-cycle one disappeared soon from the scene.

The most consistently successful developments on the Diesel engine are:

1.   The Development of the Two-Cycle Engine

The desire to reduce the cost and simplify the design soon drew attention to the two-cycle principle. From a humble beginning and a rather stepchild existence, it was developed by persistent work to a point where it holds now the leading position in the industry. It developed that the two-cycle demanded more research in the Experimental. Departments and in general more expense before it could be put on the ·market, compared with the four-cycle. It therefore had to - wait until companies with large financial and research resources got interested in the Diesel. business. They wanted to build them on a production basis and invariably selected the two-cycle. The advantages of their simplicity demonstrated themselves quickly in the field.

2.   The Tendency Toward Higher Rotative and Piston Speed

The first Diesel had a piston speed of 393.25 feet per minute. By technological progress and discarding of ideas that had outlived their use- fulness, this has been raised in our latest locomotive type O.P. engine to 1417 feet per minute. The first diesel ran at 160 RPM. There again engineering progress in metals and in lubrication have enabled us to run our locomotive engines at 850 RPM, and our smallest four-cycle engine at 1800 RPM. Hand in hand with this speeding up goes the reduction of the reciprocating weights and consideration of balance and torsional vibrations. There again the single acting engine without crosshead made the ·most progress, the engine that Fairbanks, Morse & Co. had decided from the beginning.

3.   The Supercharging of the Cylinder

A1 our engines-with separate scavenge supplied with excess air. By proper timing of the inlet and exhaust ports, this surplus air has been used in our O.P. engine to supercharge the cylinder a certain amount. The result is an MEP of 95.5 PSI.

4.   Development of the Turbocharged Engine

Where the engine rating is almost determined by the physical strength of the engine.

5.   The Developing of the Engine Into a Dual Fuel Engine

The engine can run on liquid fuel as a conventional Diesel or on gaseous fuel, using our large resources of natural gas.

In conclusion, we find that development in Diesel engineering goes ahead fast, and opinions have to be changed accordingly. What was considered a high-speed engine yesterday is a relatively slow speed one today.

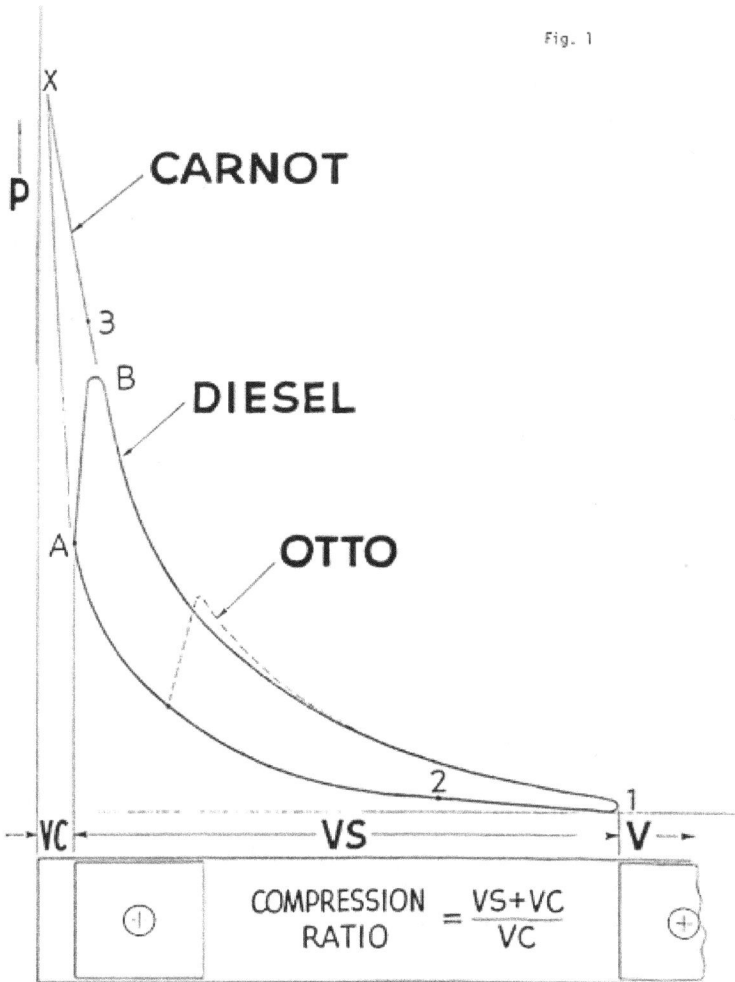

Fig. 1